Oscar Wilde's Short Stories

Adapted by **Victoria Heward**
Illustrated by **Giovanni Manna**

Editors: Claudia Fiocco, Rebecca Raynes
Design and art direction: Nadia Maestri
Computer graphics realisation: Sara Blasigh

© 2003   Black Cat

First edition: January 2003

Picture credits:
p. 5 *Oscar Wilde* Irish Tourist Board, Dublin; p. 6 *The Judge: a thing of beauty not a joy
forever*, caricature of Oscar Wilde, published in New York, 1883 Private Collection/
The Bridgeman Art Library; p. 29 *Queen Elizabeth II in her Coronation Robes*, 1953, by
C. Beaton, Victoria and Albert Picture Library; p. 64 *Awaiting Admission to the Casual
Ward* by Luke Fildes, Picture Collection at Royal Holloway, University of London; p. 67
Charles Dickens by Herbert Watkins, Hulton Getty Archives; p. 68 *Emmeline Pankhurst*,
Hulton Getty Archives.

We would be happy to receive your comments and suggestions,
and give you any other information concerning our material.
http://publish.commercialpress.com.hk/blackcat/

CISQ CISQ CERT
TEXTBOOKS AND
TEACHING MATERIALS
The quality of the publisher's
design, production and sales processes has
been certified to the standard of
UNI EN ISO 9001

ISBN 978 962 07 0498 7 Book + CD

Contents

Introduction – Oscar Wilde   5

THE YOUNG KING   7

PART ONE   The Old King's Secret   10
UNDERSTANDING THE TEXT   14

PART TWO   The Dreams   15
UNDERSTANDING THE TEXT   23

PART THREE   The Coronation   24
UNDERSTANDING THE TEXT   28

The Coronation   29
The Crown Jewels   30

THE STAR–CHILD   33

PART ONE   The Baby   37
UNDERSTANDING THE TEXT   41

PART TWO   The Mother   42
UNDERSTANDING THE TEXT   47

PART THREE   The Punishment   48
UNDERSTANDING THE TEXT   51

PART FOUR   Three Pieces of Gold   52
               UNDERSTANDING THE TEXT   61

Life in Victorian Times   64
Famous Victorians   67

THE NIGHTINGALE
AND THE ROSE

                                                             69

PART ONE   The Student in Love   72
               UNDERSTANDING THE TEXT   75

PART TWO   The Nightingale's Sacrifice   76
               UNDERSTANDING THE TEXT   79

PART THREE   The Red Rose   81
               UNDERSTANDING THE TEXT   84

PART FOUR   The Professor's Daughter   86
               UNDERSTANDING THE TEXT   89

SPECIAL FEATURES: PET Cambridge **PET**-style exercises   14, 28, 47, 63
                                                                 80, 84, 85
                      T: GRADE 5 Trinity-style exercises   32, 66, 91
                      PROJECT work using the web   32, 66
                      Exit Test – Portfolio   93

The text is recorded in full.

 These symbols indicate the beginning and end of the extracts linked to the listening activities.

Oscar Wilde.

Introduction

O scar Wilde was Irish. He was born in Dublin in 1854. His parents were very famous people. His father, William, was an important doctor. His mother was a poetess. She was called Jane but preferred the name 'Speranza'. She thought it was more interesting and romantic than Jane. Oscar's parents invited many clever and important people to their house in Dublin. They spoke together about clever and important things. When Oscar was a young boy he loved listening to them.

He studied at Oxford University and won prizes for his poetry. Oscar Wilde was a very good writer but he preferred talking.

He was also a very funny and clever man. People wanted to listen to him and to laugh at his jokes. He was very popular and everybody invited him to their dinner parties.

He wrote poems, short stories, plays for the theatre and one novel. Two of his most famous works are his novel *The Picture of Dorian Gray* and the play *The Importance of Being Earnest*.

The Judge: a thing of beauty not a joy forever,
caricature of Oscar Wilde, published in New York, 1883.

THE YOUNG KING

BEFORE YOU READ

 Here are the names of some precious objects in the story.
What colour are they?

a. ivory   **b.** amber   **c.** jade   **d.** turquoise   **e.** ruby

....

2 Connect these people to the correct definition.

a. This person is not free and must work for other people.

b. This very poor person asks others for money.

c. This person travels around the world to sell and buy things.

d. He is a very important man in the church.

e. This person looks after sheep and goats.

2. ☐ shepherd

1. ☐ bishop

3. ☐ merchant

4. ☐ beggar

5. ☐ slave

8

3 Here are some more words from the story. Connect the word to the correct picture.

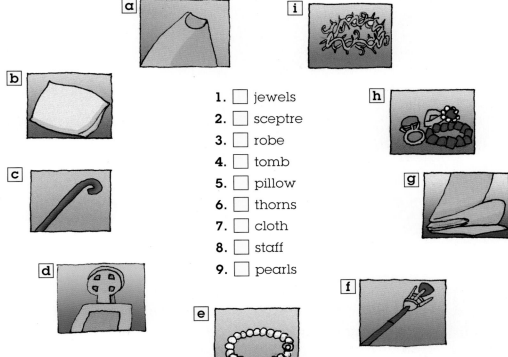

1. ☐ jewels
2. ☐ sceptre
3. ☐ robe
4. ☐ tomb
5. ☐ pillow
6. ☐ thorns
7. ☐ cloth
8. ☐ staff
9. ☐ pearls

4 Here are some verbs from the story. Do you know them?
Find the past tense of the verb and write it under the infinitive.
Then put the correct verb into the sentences.

dig   grow   throw   laugh   kneel
....

a. Tommy .... the ball and his dog ran to get it.

b. Amanda .... 10 centimetres last year!

c. The children .... because the joke was very funny.

d. The pirate .... a big hole because he wanted to find some treasure.

e. When I met the Bishop, I .... in front of him.

9

PART ONE

The Old King's Secret

here was once an old King. He had no son to be King when he died. His people were very worried. 'Who will be the next King?' they asked. But before the King died he revealed a secret: his only daughter, the Princess, had a child. In secret, she married an ordinary man and they had a son. Some people said her husband was an artist and some people said he was a musician. But his identity was a mystery and nobody knew about their secret son.

When the baby was a week old some men took him while his mother was sleeping. The Princess died immediately. Some people said for sadness. Other people said someone gave her poison [1] in a cup of wine. The men left the baby with a very poor family. This poor family lived in the forest and the boy became a shepherd. He looked after goats all day.

---

1. **poison** : a substance that causes illness or death.

THE YOUNG KING

The old King made an important decision: 'The boy must be the new King when I die,' he said. He sent his servants into the forest. 'Find the boy and bring him here.'

The servants found the boy and brought him to the palace. When the boy arrived he was very happy. He immediately fell in love with all the beautiful things around him. He took off his old leather tunic and put on his fine new clothes. Then he began to explore the castle. He ran from room to room admiring all the beautiful statues, paintings and jewels in the palace. The people of the city talked about him: 'The young King spends all his time admiring statues,' they said. 'Beauty and art are the most important things for him.' In fact the young King was so fascinated with beautiful objects that he wanted more of them. He sent merchants to India to buy ivory and jade. He sent men to Persia for silk carpets, and others to find amber in the north. He sent servants to look for green turquoise in the magic tombs of the Egyptian kings.

The young King thought about all these things but most of all he thought about his coronation robe.

He was sixteen and it was his coronation day the next day. He was very happy because he had a beautiful coronation robe of gold, a crown of rubies and a sceptre of pearls. He ordered men to work night and day to prepare his coronation robes. 'Search the whole world for the biggest rubies for my crown and the most beautiful pearl for my sceptre,' he said.

The young King was in his beautiful bedroom and he was thinking about his coronation robes. It was night time and he looked around him. The room was full of silver and gold and beautiful colours. Through the window he could smell the perfume of jasmine, [1] he could hear a nightingale singing and he could see the moon shining. Servants arrived and put flower petals on his pillow. He was very happy. Tomorrow was his coronation day. He played beautiful music on his lute and at midnight he closed his eyes and went to sleep.

That night the young King had a dream.

1. **jasmine** : a type of flower.

UNDERSTANDING THE TEXT

 Answer the following questions.

    a. Who was the princess's husband?

    b. What happened to the baby after he was born?

    c. Why was the boy happy when he arrived at the palace?

    d. What did the people of the city think of the young King?

    e. Why did the young King send his servants around the world?

    f. Describe the young King's bedroom.

PET **The King is very old and everyone is worried. What will happen when he dies? Look at the questions below. You will hear a conversation between the servant and the King. For each question, put a tick (✔) in the correct box.**

1. The old King is very
    ☐ **A** tired   ☐ **B** old   ☐ **C** worried

2. The King had a
    ☐ **A** son   ☐ **B** daughter   ☐ **C** secret servant

3. The servant thinks that a princess couldn't be a
    ☐ **A** shepherd   ☐ **B** problem   ☐ **C** king

4. The princess had a secret
    ☐ **A** son   ☐ **B** servant   ☐ **C** boyfriend

5. When they hear the secret everybody is
    ☐ **A** worried   ☐ **B** angry   ☐ **C** surprised

6. The King tells his servant to
    ☐ **A** go away   ☐ **B** bring the boy to the castle   ☐ **C** find the princess

7. The boy will
    ☐ **A** live in the forest   ☐ **B** have lots of goats   ☐ **C** be the new King

8. The servant thinks
    ☐ **A** it's a good idea   ☐ **B** the boy is not a good choice   ☐ **C** the King is tired

PART TWO

The Dreams

his was the young King's dream:

He is in a horrible, dark building. There is a terrible smell, the small windows have bars and not much sunlight can enter. But in the poor light the young King sees lots of people working. They are making cloth. They are very thin. Their faces are hungry and their hands tremble because they are tired. Pale, ill children sit in the dark corners of the room. The young King watches them.

A man speaks to him angrily and asks, 'Why are you watching me? Are you my master's spy?'

'Who is your master?' asks the young King.

'A man like me, but I have poor clothes and I am very hungry. He wears beautiful clothes and is very rich. We work for him all day. We make wine and he drinks it. We work on the land but he eats the food. We are his slaves.'

15

THE YOUNG KING

'But this is a free land,' says the young King. 'You are no man's slave.'

'In war, weak men are the slaves of strong men. In peace, poor men are the slaves of rich men.'

'Are you all slaves?'

'Yes, the women and the children. The old and the young.'

Suddenly the young King sees the cloth on the machine. It is gold. The young King is terrified. 'You are making some very beautiful gold cloth. What is it?'

'It is for the coronation robe of the young King,' the man replies.

When the young King heard this he screamed and woke up. But then he saw the yellow moon at the window and soon slept again. He had another dream.

THE YOUNG KING

This was his second dream:

He is on a long boat. The sun is very hot and a hundred slaves are rowing [1] the boat and working. The master of the boat is giving orders: he is black like ebony and has a red silk [2] turban on his head. In his ears he has big silver earrings. Someone is whipping [3] the slaves while they work.

Finally the boat arrives in a small bay and the master throws the anchor and a long rope ladder [4] into the sea. Some men take the youngest slave. They tie a heavy stone to him and throw him into the sea. The young slave disappears into the water but returns to the boat many times. Every time he returns he has a beautiful pearl in his hand. The master of the ship looks at the pearls and puts them into a little green bag.

The young slave returns to the boat for the last time. He is very pale and tired. In his hand he has a very beautiful pearl. The pearl is round and white like the moon. But the young slave's ears and nose are full of blood. He falls and dies but the master of the boat laughs. He takes the pearl from the young slave's hand and the other slaves throw his dead body into the sea. 'This pearl is for the sceptre of the young King,' he says.

When the young King heard this he screamed and woke up but he saw the stars at the window and soon slept again.

---

1. **rowing** : making the boat move in the water with oars.
2. **silk** : a fine cloth.
3. **whipping** : hitting with a whip.
4. **rope ladder** :

The Dreams

This was his third and last dream:

He is in a tropical forest. It is full of strange fruit and beautiful, poisonous [1] flowers. There are snakes in the grass, parrots [2] in the trees and monkeys [3] and peacocks [4] all around. The young King sees lots of men working in a dry river. They are digging the ground and cutting big rocks and stones.

Death [5] and Avarice [6] are in a dark cavern. They are also watching the men. Death says to Avarice, 'Give me one third of your men,' but Avarice refuses. 'No! They are my servants,' she says.

Death is very angry when he hears this. He sends Malaria to kill one third of the men.

'What have you got in your hand?' Death asks.

'Three grains of corn. [7] But why are you interested?' she asks.

Death says, 'Give me one grain of corn to plant in my garden.' But Avarice replies, 'No, it is my corn,' and she hides the corn in her pocket.

Again, Death is very angry when he hears this and calls Fever. [8] Fever comes in a red robe like fire, he touches one third

1. **poisonous** : containing a substance that can kill you.
2. **parrots** : type of tropical bird.
3. **monkeys** :
4. **peacocks** :
5. **death** :
6. **avarice** : extreme selfish desire for something.
7. **grains of corn** :
8. **fever** : if you have a fever your body temperature is high.

of the men and kills them. 'Now give me a grain of corn for my garden,' says Death.

'No, never!' replies Avarice. Death is extremely angry and calls Plague. [1] Plague arrives from the sky, flying like a bird and kills the rest of the men. Avarice screams and runs into the forest. Death takes his red horse and rides away, fast like the wind. And then dragons and terrible monsters come out of the rivers and the valleys.

The young King cries and says, 'Who were those men? What were they doing?'

'They were looking for rubies for a king's crown,' replies a voice behind him. The young King turns and sees a man in white. This man has a mirror in his hand.

'Which king?' he asks.

'Look in this mirror and you will see the king,' replies the man in white.

He sees his face in the mirror, screams and wakes up. He sees the sun shining at his window. It is his coronation day.

---

1. **plague** : contagious disease which kills many people.

UNDERSTANDING THE TEXT

1 **Write questions for these answers.**

FIRST DREAM

a. ....? It is horrible and very dark.

b. ....? They are making cloth.

c. ....? No. Everybody is a slave.

SECOND DREAM

d. ....? They are rowing the boat and working.

e. ....
........? He throws the anchor and a long rope ladder into the sea.

f. ....? Into a little green bag.

THIRD DREAM

g. ....? Snakes, parrots, peacocks and monkeys.

h. ....? In a dark cavern.

i. ....? Malaria, Fever and Plague.

2 **Match each sentence on the left to the words on the right. There are two extra answers which you don't need to use.**

1. ☐ Their hands tremble.	a. Death
2. ☐ They sit in dark corners.	b. The children
3. ☐ He eats good food and drinks good wine.	c. Avarice
4. ☐ He wears a red turban.	d. Plague
5. ☐ They tie a heavy stone to him.	e. The master of the ship
6. ☐ They are in a dark cavern.	f. Fever
7. ☐ He wears a red cloak.	g. The young King
8. ☐ He flies like a bird.	h. Dragons and monsters
9. ☐ He wants one grain of corn.	i. Death and Avarice
10. ☐ She hides the corn in her pocket.	j. The youngest slave
	k. The machine workers
	l. The master of the machine workers

23

PART THREE

The Coronation

 servant arrived with the coronation clothes. They were extremely beautiful but the young King remembered his dreams. 'Take these clothes away. I don't want to wear them,' he said.

'Is this a joke, Your Majesty?' asked the servant, but the young King told him about his dreams.

'In my robe there is sadness and pain, in the rubies there is blood and in the pearl there is death,' he said.

The servant replied, 'Please forget your dreams. Put on the robe and the crown. The people will not recognise a king without a crown and a sceptre.'

But the young King put on his old tunic from the forest and took his shepherd's staff. 'I arrived in the palace with these clothes and I will leave the palace with these clothes,' he said. 'Now I am ready for my coronation.'

The Coronation

A servant asked him, 'Where is your crown?' And he took a briar [1] of thorns from his balcony. 'This will be my crown,' he replied.

The young King rode his horse to the cathedral. The people laughed when they saw him. 'This is not the King but the King's servant,' they said. He explained his dreams but one man was angry and said, 'Do you not know that rich people give poor people work. It is difficult to work for a hard master but it is more difficult to work for no master. Please return to the palace and put on your coronation robes.'

'The rich and the poor are brothers,' he replied, but the people laughed again.

He arrived at the great door of the cathedral but the soldiers stopped him. 'What do you want? Only the King can enter by this door.'

'I am the King,' he replied. The Bishop saw him and asked, 'Where is your crown? Where is your sceptre?'

The young King told the Bishop of his dreams but the Bishop answered, 'Listen to me, I am an old man. There are many bad things in the world but you cannot change them all. There are thieves and pirates and beggars but you can't make these things disappear. They are too much for one person. Go back to the Palace and put on your coronation clothes.'

But the young King passed the Bishop and entered the cathedral. He went to the altar and looked at the image of Christ. He saw the light of the candles and the smoke of the

---

1. **briar** : a wild rose with long, thorny stems.

THE YOUNG KING

incense. Suddenly a crowd of people ran into the cathedral. They had swords and were very angry. 'Where is this King dressed in beggar's clothes?' they cried. 'We must kill him because a beggar cannot rule us. He will be bad for our country.' But the young King prayed silently in front of the altar. Then he turned and looked at the people sadly.

At that moment a ray of sun shone into the cathedral. It illuminated the young King at the altar. The sun made a beautiful robe around him, red roses grew on his dry crown of thorns and white lilies grew on his staff. The roses were redder than rubies and the lilies were whiter than pearls. Music started to play and voices started to sing. The glory of God filled the cathedral. The people knelt down.

'He is crowned [1] by someone greater than me,' the Bishop said and he knelt in front of the young King. The boy came from the altar and passed the people. But they didn't have the courage to look at his face because it was the face of an angel.

1. **crowned** : officially declared king.

UNDERSTANDING THE TEXT

PET 1 **Look at the sentences below about the whole story. Decide if each sentence is correct or incorrect. If it is correct mark A, if it is not correct mark B.**

		A	B
1.	The young King finds beautiful jewels in India.	☐	☐
2.	Servants put gold on his pillow.	☐	☐
3.	The young King has three dreams.	☐	☐
4.	The man in the first dream has a bad master.	☐	☐
5.	In the second dream the ship's master gives the young King a pearl.	☐	☐
6.	Death wants a lot of corn from Avarice.	☐	☐
7.	The young King doesn't want to wear his coronation robe.	☐	☐
8.	The young King says the rich are more important than the poor.	☐	☐
9.	The people want their king to wear a real crown.	☐	☐
10.	The young King is crowned by God.	☐	☐

2 **You will hear Max and Julia speaking. They are watching the coronation procession of the young King.**
Listen to their conversation and look at the summary below. Some information is missing. Fill in the missing information in the numbered spaces.

The young King arrives. He is wearing a [1].................... of briars and a leather [2].................... .
This is strange because a king usually has a golden [3].................... .
Max asks the King why he is wearing the clothes of a [4].................... man.
The young King thinks that [5].................... are important.
'The [6].................... and the poor are [7]....................,' he says.
[8].................... agrees with the King but [9].................... thinks a king must wear a king's [10].................... .

THE CORONATION

Here is a picture of Queen's Elizabeth's coronation in 1953. She is wearing a crown and holding a sceptre and an orb[1] in her hand. All of these things are called 'The Crown Jewels'. She is wearing a beautiful robe and sitting on the coronation throne.

Queen Elizabeth II in her Coronation Robes, 1953, by C. Beaton.

Victoria and Albert Picture Library

1. **orb** : globe.

In 1066 a French nobleman called William, went to England. He defeated the English and became the new King. Everyone called him 'William the Conqueror' and he was the first king to be crowned in Westminster Abbey. Now all kings and queens are crowned here.

A coronation is always a moment for celebration but it is also a religious ceremony. British monarchs must promise to govern with justice and mercy [1] when the Archbishop of Canterbury [2] puts the crown on their head.

THE CROWN JEWELS

This is St. Edward's Crown. It was made in 1661. All new kings and queens of England wear it at their coronations. Which jewels can you see in it?

The orb is made of gold and was specially designed for the coronation of King Charles II in 1662. The monarch holds it in his or her right hand. It shows that the king or queen is a Christian. The sword represents authority and the ring represents dignity.

1. **mercy** : compassion.
2. **Archbishop of Canterbury** : the head of the Church of England.

Did you know?
Some interesting facts about the Crown Jewels

- In 1216 King John loses the Crown Jewels in some quicksand. [1]
- Edward III pawns [2] the jewels to make money to pay his soldiers.
- Oliver Cromwell orders them to be destroyed in 1649 because they are a symbol of the monarchy.
- The state crown contains the famous Koh-i-noor diamond. It is one of the biggest diamonds in the world and travels from India to Afghanistan to Persia before it arrives in England as a gift for Queen Victoria.

What do you think?

1 **The young King has a golden robe, a crown of rubies and a sceptre of pearls. What do you think these things represent? Choose from this list. Add your own ideas if you want.**

☐ purity   ☐ authority

☐ power   ☐ blood

☐ money   ☐ justice

☐ flowers   ☐ beauty

2 **But later he decides to wear his leather tunic, a crown of thorns and to carry a shepherd's staff. Do you think these things represent something different?**

1. **quicksand** : deep, wet sand which sucks in anyone or anything on it.
2. **pawns** : if you pawn something, you leave it in possession of a person (a pawnbroker) who will give you money for it and will keep it until you can pay back the money.

PROJECT ON THE WEB

MONARCHIES AROUND THE WORLD

1 Work in groups to find information about different monarchies around the world.
Use one of the following search engines to help you:

http://www.altavista.com
http://www.yahoo.com

2 With your group decide what you want to research and type in:
'Monarchy'
'British Royal family'
'Monarchies around the world'
'Monarchy of…. (name of country)'

3 Make presentations to show the others in your class what you have found. Why not display your work on the classroom wall?

T: GRADE 5

 Topic – Celebrations
Find some information/a picture about a celebration such as a royal wedding, if possible from your country.
Tell the class about the celebration using these questions to help you:

a. Where and when was the celebration?

b. Who attended the celebration and what did they do to celebrate?

c. Describe the clothes, food and music.

d. Have you been to a celebration such as a wedding, a carnival, or a special birthday? Can you describe it?

THE STAR-CHILD

BEFORE YOU READ

1 These words from the story are connected to Nature.

   a. Match the words to the pictures.
   b. Put them into the correct column in the table on page 35.
   c. Add three more words to each column.

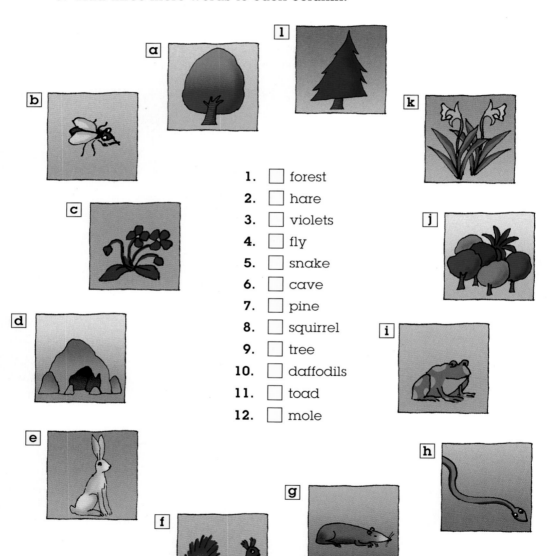

1. ☐ forest
2. ☐ hare
3. ☐ violets
4. ☐ fly
5. ☐ snake
6. ☐ cave
7. ☐ pine
8. ☐ squirrel
9. ☐ tree
10. ☐ daffodils
11. ☐ toad
12. ☐ mole

Animals	Reptiles and insects	Places	Plants and flowers
....	....	....	....
....	....	....	....
....	....	....	....
....	....	....	....
....	....	....	....
....	....	....	....
....	....	....	....

2 **Circle the best alternative.**

a. Aeroplanes **fly** / **kneel** in the sky.

b. Babies **fly** /**cry** when they are hungry.

c. Friends **beat** / **hug** each other when they are happy.

d. Detectives **kiss** / **follow** suspects.

3 **Connect the word to the statement.**

1. ☐ hunger
2. ☐ punishment
3. ☐ prophecy
4. ☐ hatred

5. ☐ wonder

6. ☐ pity

7. ☐ mercy

a. 'I hate snakes and spiders.'

b. 'Wow!!!'

c. 'I haven't got any food to eat.'

d. 'You are a bad man, but you will not go to prison. I will give you a second chance.'

e. 'In the year 2012 you will become very rich.'

f. 'You are a very bad dog. No ice-cream for you today!'

g. 'You have more problems than me. I will help you.'

4 Connect the word to the picture.

a. pot	**b.** trap	**c.** chain
d. chest	**e.** crowd	**f.** wings

1 ☐
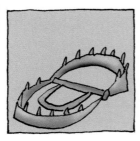

2 ☐

3 ☐

4 ☐

5 ☐

6 ☐
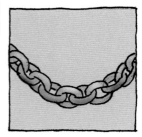

5 Look at the picture on page 39.

a. With your partner describe the scene.

b. Who do you think the two people are?

c. What do you think they are feeling?

d. What do you think will happen?

PART ONE

The Baby

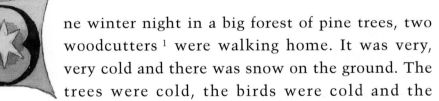

One winter night in a big forest of pine trees, two woodcutters [1] were walking home. It was very, very cold and there was snow on the ground. The trees were cold, the birds were cold and the animals were cold. The rabbits stayed in their rabbit holes and the squirrels stayed in the trees.

But the two woodcutters continued their journey. They prayed to Saint Martin the protector of travellers and finally they saw the lights of their little village in the distance. They were very happy and laughed. The Earth now seemed like a flower of silver and the moon seemed like a flower of gold. But soon they became sad again. 'Why were we so happy?' asked one woodcutter. 'Life is for rich people, not poor people like us. It is better if we die in the snow or if a wild animal eats us.'

---

1. **woodcutters** : people who cut down trees as a job.

THE STAR-CHILD

Suddenly, something very strange happened. A very bright and beautiful star fell from the sky into the snow.

'Look,' said one of the woodcutters to his friend, 'perhaps we will find a pot of gold. Let's go and see!'

When they arrived they found a thing of gold on the white snow. But it wasn't the treasure they wanted. It was a golden cloak with golden stars on it. They opened the cloak and inside they saw a little baby, sleeping. Round the baby's neck was a chain of amber.

'This is not good,' said one of them. 'Let's leave the baby here. We have too many children and not enough money to buy food. I don't want another child.'

'But we can't leave this little baby here alone,' said the other woodcutter. 'He will surely die. I will take him home with me. We have many children and not enough food, but my wife will look after him.' And the good woodcutter took the baby in his arms and continued his journey home.

When they arrived at their village the first woodcutter said, 'You have the child so you must give me the cloak of gold.'

But his friend answered, 'No, this cloak is not yours or mine. It is the baby's cloak. It must stay with him.'

The woodcutter's wife was very happy to see her husband. She put her arms round him and kissed him.

'I found something in the forest and I brought it home for you,' he said.

'Good, what is it? We are very poor and we need many things.'

But she was very angry when she saw the baby. 'We have too many children already and not enough money to buy food. I don't want another child,' she said. But then she looked at the baby

THE STAR-CHILD

and her heart was full of pity.

'He is a Star-Child,' said her husband. 'We must love him.'

So the woman put the baby in a little bed to sleep. She put the cloak and the chain into a chest. 'Yes, we will love him,' she answered.

UNDERSTANDING THE TEXT

1 **Complete this summary of Part One. Write one word in each space.**

Two ¹................. are walking home. It is winter and ²................. are
tired and very cold. Suddenly they see something fall out of the
³................. . They go to see what it is and they find a baby: a
⁴.................-Child! The baby has a ⁵................. cloak and a chain of
amber round his ⁶................. . One of the woodcutters decides to take
the ⁷................. with him. When he arrives home his ⁸................. is
very angry because they have too many ⁹................. and not enough
¹⁰................. to feed them. But then she feels ¹¹................. for the
baby and puts him in a ¹²................. to sleep and decides to love him.

2 **The first woodcutter took the baby home to his wife. The second
woodcutter went home alone. Imagine you are the second
woodcutter and invent the answers to these questions.**

Name...   Age...   Wife's name...

**Now work with a partner: talk about these questions in your
language and then take turns to be the second woodcutter and a
journalist. Complete the interview and then act it out.**

Journalist: Hello... (name of woodcutter), I understand something very
strange happened to you today. Is it OK if I ask you some questions?

Woodcutter: Yes, of course.

Journalist: First, tell me what happened when you were coming home in
the forest.

Woodcutter: Well, my friend and I were walking home. It was very cold
and suddenly...

Journalist: Very interesting and what do you think about the baby, the
chain of amber and the cloak of gold?

Woodcutter: I think...

Journalist: What do you think of your friend's decision? Do you agree with
him? Did he do the right thing

Woodcutter: Hmmm. That's a difficult question. I think...

Journalist: What will you tell your wife when you get home? Will she
be happy or angry with you?

Woodcutter: My wife will...

PART TWO

The Mother

he Star-Child lived with the woodcutter and his family but he was very different from them. Every year the Star-Child became more beautiful: his skin was white like ivory, his hair was gold like the daffodils, his lips were like the petals of a red flower and his eyes were blue like the violets near a river. The other people in the village had black hair and black eyes and they watched the Star-Child in wonder.

The Star-Child was very beautiful but very cruel, arrogant and selfish. He laughed at the other children in the village and said, 'Your parents are poor but I am noble, I come from a star.' He had no pity for poor people. He laughed at ugly people and ill people. He hurt [1] animals and he laughed when they suffered.

---

1. **hurt** : gave pain to.

He was very vain and loved his beauty. In summer he often went to the well [1] in the priest's orchard and looked at the reflection of his face in the water. Then he was happy.

The woodcutter and his wife treated the boy well but they were very sad. They often said to him, 'We were good to you. We felt pity for you. Why are you so cruel? Why do you act in this way?'

The priest was very worried and said to him, 'You must respect all God's creatures. Even the fly is your brother. Why do you cause pain to others?'

But the Star-Child didn't listen. He continued to hurt animals and laugh at the problems of other people. The other children followed him because he was beautiful and could dance and make music. They followed his orders. He was their leader and they became cruel and hard like him.

One day a poor beggar woman arrived in the village. Her clothes were very old and torn and she had no shoes on her feet. She was very tired and sat under a tree to rest. The Star-Child saw her and said to his friends, 'Look at that ugly woman. We don't want her here,' and they started to throw stones at the poor woman. She was terrified but she didn't stop looking at the Star-Child.

'What are you doing?' shouted the woodcutter when he saw this. 'Stop immediately. Why do you have no pity for this poor woman?'

1. well :

THE STAR-CHILD

'I will not listen to you. You are not my father,' replied the Star-Child.

'This is true, but when I found you in the forest I had pity for you.'

The old woman was listening and when she heard these words she screamed and fainted. [1] The woodcutter carried her into his house and his wife put meat and drink on the table for her. But she did not eat or drink. She asked, 'Did this child come from the forest? Did he have a gold cloak with stars on it? Did this happen about ten years ago?'

The woodcutter was very surprised. 'Yes,' he replied.

'And did he have an amber chain round his neck?'

'Yes he did,' said the woodcutter. 'Come with me and I will show you the cloak and the chain.'

The woman looked at these things and started to cry with joy. 'He is my little son. I am his mother,' she said. 'I lost him in the forest ten years ago and I looked all over the world for him. Now I have him again.' The woodcutter was very surprised and called the boy. 'Come into the house and you will find your mother.'

The Star-Child was very happy and ran in but when he saw her he said, 'Where is my mother? I can see no-one, only a horrible beggar woman.'

'I am your mother,' she said.

'You are mad. I am not your son: you are dressed in old clothes, you are a beggar woman and I am a Star-Child!'

'But I recognised you when I saw you and I recognised your

---

1. **fainted** : lost consciousness.

THE STAR-CHILD

cloak of gold and your chain of amber. Robbers stole you from me. Come to me, my son. Your love is very important for me.' She opened her arms to him but he was very angry and closed the doors of his heart to her.

The woman cried. 'Kiss me before I go because I travelled all over the world and I suffered much to find you.'

'Never. You are very ugly. I prefer to kiss a toad or a snake.'

The woman stood up and went out of the house. She was crying very much. The Star-Child was very happy when she went. He then went to play with his friends.

UNDERSTANDING THE TEXT

**Choose the correct words to complete the sentences.
Tick A, B, C or D.**

1. The Star-Child has
   - ☐ **A** black hair and blue eyes
   - ☐ **B** blonde hair and white skin
   - ☐ **C** black hair and blue skin
   - ☐ **D** blonde hair and black eyes

2. The woodcutter and his wife
   - ☐ **A** are happy with the boy
   - ☐ **B** don't understand why the boy is cruel
   - ☐ **C** think the boy is funny
   - ☐ **D** love to play with the boy

3. The priest
   - ☐ **A** thinks the Star-Child is beautiful
   - ☐ **B** laughs at the problems
   - ☐ **C** loves the Star-Child
   - ☐ **D** says that all animals are men's brothers

4. The other children
   - ☐ **A** love to play with the animals
   - ☐ **B** listen to the priest
   - ☐ **C** don't like the Star-Child
   - ☐ **D** become like the Star-Child

5. The old woman
   - ☐ **A** throws stones at the Star-Child
   - ☐ **B** has no pity for the woodcutter
   - ☐ **C** is very poor and tired
   - ☐ **D** wants the chain of amber

6. She knows the Star-Child is her son
   - ☐ **A** when she hears the woodcutter's words
   - ☐ **B** when he starts to throw stones at her
   - ☐ **C** because she knows his address
   - ☐ **D** because he says 'Hello Mother'

7. The Star-Child
   - ☐ **A** wants to kiss a snake
   - ☐ **B** thinks the woman is very ugly
   - ☐ **C** is happy to meet his mother
   - ☐ **D** recognises the woman immediately

PART THREE

The Punishment

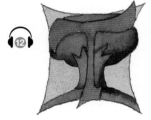

he Star-Child went to his friends but when they saw him they laughed at him. 'We don't want to play with you now because you're ugly,' they said.

'Why do they say these things to me?' he thought. He went to the well to look at his reflection. He was different now: he had a face like a toad and skin like a snake.

Then the Star-Child understood and started to cry. 'This is my punishment,' he said. 'I am very cruel and my mother suffered. Now I must find her and say sorry.'

The woodcutter's little daughter said to him, 'It's not important if you're ugly. Please stay. I will not laugh at you.'

'No, this is my punishment,' he replied. 'I treated my mother very badly and now I must find her.'

He ran into the forest calling, 'Mother! Mother! I'm sorry, please come back.' All day he called but nobody answered. When night came he slept on a bed of leaves, but when the animals saw

him they ran away. They knew that he was a cruel boy.

He said to the mole, 'You can go under the ground. Tell me if my mother is there?'

'I don't know if your mother is there. I cannot see because you hurt my eyes,' replied the mole.

He said to a little bird, 'You can fly over the trees. Tell me if you can see my mother.'

'I don't know if your mother is there. I cannot fly because you hurt my wings,' replied the bird.

He saw a little squirrel and asked, 'Where is my mother?'

'I don't know,' replied the squirrel. 'You killed my mother. Do you want to kill your mother too?'

The Star-Child heard all these things and he cried and prayed to God to forgive him. He travelled to many different villages to find his mother, and the children of these villages laughed at him and threw stones at him. Nobody had pity for the Star-Child.

UNDERSTANDING THE TEXT

 Who is speaking and who is listening?

Here's an example:

'We don't want to play with you.'

The children are speaking.

The Star-Child is listening.

a. 'This is my punishment.'

....

....

b. 'Please stay. I will not laugh at you.'

....

....

c. 'You can go under the ground.'

....

....

d. 'I cannot fly because you hurt my wings.'

....

....

e. 'You killed my mother.'

....

....

2 **Find a word in Part Three which means:**

a. A hole in the ground with water in it.   ....

b. This covers all of your body.   ....

c. They are usually green and grow on trees.   ....

d. A small animal who lives under the ground   ....
and cannot see very well.

e. Small towns in the countryside.   ....

PART FOUR

Three Pieces of Gold

or three years the Star-Child walked around the world but he didn't find his mother. One day he arrived at the gates of a city near a river with a big wall around it. The soldiers there stopped him. 'What are you doing here?' they asked.

'I'm looking for my mother,' he said. 'Please let me pass. Perhaps she is in this city.'

'Who is your mother and why are you looking for her?' asked another soldier.

'She is a poor beggar like me and I was very cruel to her. Now I want her pardon.'

But the soldiers laughed. 'You are very ugly. No mother loves an ugly child. She will not be happy to see you. Come with us. We will sell you to be a slave.'

They sold the Star-Child to an old man for the price of a cup of

Three Pieces of Gold

sweet wine. This old man was a magician [1] from Libya. The Magician took the boy to a dark prison and gave him a piece of old bread and some dirty water. The next day he said, 'Now you must go into the forest. In the forest there are three pieces of gold: one is of white gold, one is of yellow gold and the other is of red gold. You are my slave and if you do not bring me the piece of white gold I will beat you one hundred times.'

So the Star-Child went to the forest to look for the white gold but he found only many thorns [2] and dangerous plants. He could not find the white gold anywhere. When the sun started to disappear the boy started to cry. He knew that the Magician wanted to beat him. Suddenly he heard a cry of pain and saw a little hare in a trap. He forgot his problems. He felt pity for the hare and opened the trap.

'Thank you, you are very kind,' said the hare. 'Thanks to you I have my freedom. What can I give you?'

'I must find a piece of white gold for the Magician. If I don't take it to him he will beat me.'

'I will help you,' said the hare. 'I know where to find the white gold.' He took the Star-Child to a tree and in the tree he found the gold. The Star-Child was very happy and thanked the hare.

He returned to the city. But at the city gate he saw an old man. This old man was very ill and very poor.

'Give me some money. If you don't give me some money I will die of hunger!' shouted the old man. The Star-Child felt pity for

1. **magician** : a person who has magic powers.

2. **thorns** :

53

THE STAR-CHILD

the old man but he only had the piece of white gold for the Magician. 'The old man needs the money more than me,' thought the Star-Child and gave him the gold.

The Magician was very angry when he saw that the Star-Child didn't have the gold and he beat the boy. He put him in prison with no food and no water.

The next day the Magician said, 'Today you must return to the forest and find the piece of yellow gold. If you do not do this I will beat you three hundred times.' The boy went into the forest and looked for the gold. He looked all day long but he could not find it. Finally he sat under a tree and started to cry. The hare heard him and asked, 'Why are you crying?'

'I must find the piece of yellow gold. If I don't find it the Magician will beat me.'

'Follow me. I will show you the yellow gold,' said the hare and he took the Star-Child to a pool[1] of water. At the bottom of this pool he found the piece of yellow gold.

The Star-Child returned to the city but at the city gate he saw the old man again. 'Give me some money. If you don't give me money I will die of hunger!' he shouted. The Star-Child felt pity for the old man and gave him the gold.

The Magician was very angry. 'What!? No gold? No gold, no food and no water!' He beat the Star-Child and put chains on him and put him in prison again.

The next day the Magician said, 'Today you must return to the forest and find the piece of red gold. If you find it, you will be

---

1. **pool** :

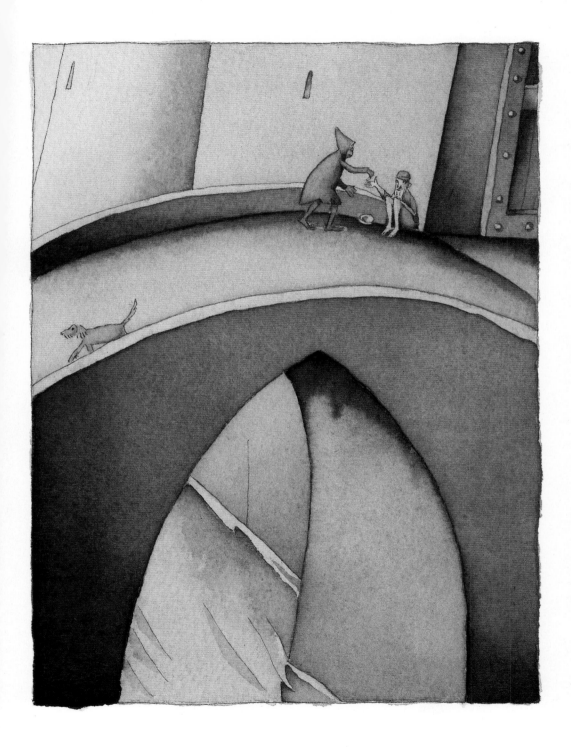

free. If you do not find it, I will kill you.' The boy went into the forest and all day long he looked for the gold but he could not find it. In the evening he sat under a tree and started to cry. The hare heard him and asked, 'Why are you crying?' The Star-Child explained everything and again the hare helped him. This time he found the gold in a cave near the tree.

'Thank you, thank you,' said the boy and he ran back to the city.

At the city gate he saw the old man. 'Give me some money. If you don't give me money, I will die!' he shouted. The Star-Child felt pity for the old man and gave him the gold. 'You need it more than me,' he said, but he was very sad and his heart was very heavy. 'The Magician will kill me,' he thought.

But when he passed the guards at the city gates they bowed[1] to him and said, 'Look at our beautiful Lord!' The Star-Child walked through the city and more and more people followed him. They all said, 'He is the most beautiful boy in the world.' But the Star-Child was very sad, 'They are laughing at me,' he thought. He walked for a long time and finally arrived in a big square where there was a king's palace. The people said, 'You are our Lord, the son of our king!'

'I am not a king's son. I am the son of a poor beggar woman. Why do you say that I am beautiful? I know I am very ugly.'

'Why do you say that you are ugly? Look!' said a soldier. The Star-Child looked into the soldier's shield.[2] The shield was silver

---

1. **bowed** : inclined their heads to show respect.

2. **shield** :

Three Pieces of Gold

like a mirror. There he saw his face and saw that his face was beautiful like before.

'There is a prophecy,' said the people, 'that on this day our king will come. You are our king. Take this crown and this sceptre. [1] Govern us with justice and with mercy.'

'No, I am a bad boy,' he replied. 'I must find my mother, I cannot accept the crown and the sceptre.'

He turned towards the city gate. In the crowd he saw his mother, the beggar woman. Then next to her he saw the old man from the city gate.

He ran to the woman, knelt in front of her and kissed her feet.

'Mother I am so sorry. Please forgive me. Once I gave you my hatred. Please give me your love now.'

But the woman didn't speak. The Star-Child spoke to the old man. 'Please, I helped you three times. Please tell my mother to speak to me.'

But the old man didn't speak. The Star-Child started to cry. 'Please forgive me, Mother. Please forgive me.'

The woman put her hand on the boy's head and said, 'Stand up.' The old man put his hand on the boy's head too. When the Star-Child stood up he saw that the beggar woman was a queen and the old man was a king. The queen said to him, 'This is your father. You helped him three times.' And the king said to him. 'This is your mother. You washed her feet with your tears.'

The boy hugged [2] them and kissed them both. They took him

---

1. **sceptre** :
2. **hugged** : put his arms around.

to the palace and they put a crown on his head and a sceptre in his hand.

He was a very good king and showed justice and mercy to everyone. He sent gifts to the woodcutter and his family. He helped poor people, he was kind to the animals and birds and there was peace in all the land.

Unfortunately he died after three years because he suffered a lot in his life and the next king was a cruel king.

UNDERSTANDING THE TEXT

 Choose the best alternative to complete the sentences in this summary of Part Four.

The Star-Child finally arrives in a [1] **village / big city**. At the gates he meets some soldiers. They tell him that [2] **mothers don't love ugly children / his mother is in the city**. The soldiers sell the Star-Child to [3] **a magician / a slave**. He takes the boy to a prison and gives him [4] **old bread / sweet wine**. The Magician promises to [5] **hit / free** him if he finds the [6] **white gold / hare**. The Star-Child sees a hare in a [7] **trap / hole** and helps him to escape. The hare helps him to find the white gold in [8] **a cave / a tree** but he gives it to [9] **a magician / an old man**. The Magician is very [10] **hungry / angry** and puts the Star-Child in prison again. The next day he finds the yellow gold [11] **in a pool / under a tree**. The red gold is [12] **at the city gate / in a cave**. When he returns to the city the [13] **people think / Magician thinks** that the boy is [14] **a soldier / a king**. The Star-Child discovers that his mother is a [15] **queen / beggar woman**. He becomes a very [16] **poor / good** king.

2 What do you think?

a. Why does the Star-Child become ugly in Part Three and beautiful again in Part Four?

b. All of these characters treated the Star-Child badly: the soldiers, the Magician, the old man, his mother. Do you agree?

c. Who treated him the worst?

d. Do you think 'The Star-Child' is a happy or a sad story? Why?

3 **Listen to this conversation twice and fill in the gaps with these words.**

> know   give   bring   early   wake up   ugly   eat
> have   generous   rude   sleep   poor

The Magician and the Star-Child are talking early one morning.

Magician: Come on boy, ¹.................. .

Star-Child: What time is it? Is it morning already?

Magician: It's 6 o'clock.

Star-Child: 6.00! That's very ².................. . Can I ³.................. some more, please?

Magician: What? This is a prison, boy, not a hotel!

Star-Child: Alright, alright. Can I have something to ⁴.................. then?

Magician: You can ⁵.................. this piece of bread.

Star-Child: Thank you very much. Very ⁶.................. . What's your name?

Magician: I haven't got a name. I'm a magician.

Star-Child: Oh!

Magician: And what's your name?

Star-Child: I haven't got a name. I'm a Star-Child.

Magician: Oh. And why are you in this city, Star-Child?

Star-Child: I'm looking for my mother. Perhaps you ⁷.................. her. She's very ⁸.................. and extremely ⁹.................. .

Magician: Like her son. Ha, ha, ha.

Star-Child: Don't be ¹⁰.................. and ¹¹.................. me some water, please.

Magician: Alright, boy, but remember, if you don't ¹².................. me the gold, there will be no more food for you tomorrow. Ha, ha, ha.

4 In the box for exercise three there are 5 adjectives. Write them in this table.

a.	....
b.	....
c.	....
d.	....
e.	....

5 Here are the opposite adjectives. Write them in the table too.

> late   rich   beautiful   mean   polite

6 Listen to the recording again. With your partner read the dialogue. Try to speak with the same voices as the Magician and the Star-Child.

7 The Star-Child and the Magician don't know each other very well. With a partner think of 3 questions the Star-Child could ask the magician.

1. ....?
2. ....?
3. ....?

Now think of 3 questions the magician could ask the Star-Child.

1. ....?
2. ....?
3. ....?

PET 8 Now imagine you don't know your partner very well. Speak together to find as much information as possible. Talk about:

name, age, family, home, school, free time

Life in Victorian Times

Oscar Wilde wrote in the second half of the nineteenth century. At this time Victoria was the Queen of England and the British Empire was the biggest and most important in the world. (Remember that Wilde was Irish not English. He went to England to study, live and work.) During this time there were many changes in the way people lived. Many inventions changed the way people worked. There was a very big difference between the rich and the poor.

England became more industrialised and people started to leave the countryside. They went to work in new factories in the cities. There were not enough houses for everybody so people had to share small houses with other large families or go to the workhouses. The workhouses were terrible places:

Awaiting Admission to the Casual Ward, by Luke Fildes.

The Picture Collection at Royal University Holloway, University of London

there was work and a place to live for poor people, but usually there was not enough food and the work was very difficult and dangerous. The factory masters and workhouse masters were usually very strict and had no pity for the workers. Lots of people died because they had accidents with the new machinery or because they worked fifteen hours every day in very bad conditions.

1 **Answer these questions.**

a. In one of the two stories you read, Oscar Wilde writes about a situation similar to a workhouse. Which story is it? What happens?

b. In the nineteenth century machines were invented to make cloth, to move engines and to do the work of many people. What are the most important modern inventions? Why?

2 **Look at this list of objects. They are all very important modern inventions.**

☐ car   ☐ aeroplane   ☐ mobile phone
☐ computer   ☐ internet   ☐ washing machine

Put them in order of importance. Number 1 is the most important and number 6 is the least important. Show your list to your partner and explain your decision. Do you agree?

3 **What invention would you like to see in the next fifty years and why? Here are some ideas: a flying car, a robot to do all the housework, a telephone video...**
Now use your imagination and think of some other ideas.

PROJECT ON THE WEB

1 In groups, choose an invention (you can use one from the list on page 65 if you want) and find some information about it. Make a wall display to show your results.

Use the Internet to help you. You can find the information you need using different search engines.

```
www.searchalot.com
www.yahoo.com
www.altavista.com
```

2 Write 'the invention of... .' and then the name of the invention you want to find. Find out when it was invented, who invented it, what it is used for etc.

T: GRADE 5

  Topic – Transport

Find a picture, photo or a timetable of a means of transport in your country.

Tell the class about it using these questions to help you.

a. How popular is this means of transport and who uses it?

b. How much does it cost and how does it affect the environment?

c. How do you and your friends travel to work/school?

d. In your opinion, what has been the best invention for travelling? Do you know who invented it and when?

Famous Victorians

🎧 ⑯ Charles Dickens (1812 - 1870)

Charles Dickens (circa 1850)
by Herbert Watkins.

Hulton Getty

We can see that Oscar Wilde thought about the differences between rich and poor people and wrote about these problems in his stories. Another very famous English writer did the same. His name was Charles Dickens. Oscar Wilde's family was very rich and very important and his stories often showed poor children who became rich by a sort of magic. Charles Dickens' family was poor and his stories were very realistic: they spoke about poor children in workhouses and bad rich people. These people were not interested in helping others. His most famous stories are *Oliver Twist*, *Great Expectations* and *David Copperfield*.

Doctor Thomas Barnardo (1845 - 1905)

Doctor Barnardo was another famous Victorian. He was Irish like Oscar Wilde. When he visited England he was very shocked by the terrible conditions of the poor, especially the children: they had to carry very heavy sacks of coal, work for many hours every day, go into chimneys to clean them, and so on. He built special houses for poor children. In these houses they could live protected and safe.

Emmeline Pankhurst (1858 - 1928)

Emmeline Pankhurst in New York (circa 1911).

Hulton Getty

The difference between the rich and the poor was not the only difference in Victorian society. There was also the difference between men and women. Women couldn't vote at that time. Men thought that they were not very intelligent and could not make important decisions! Emmeline Pankhurst was the leader of a group of women called the Suffragettes. The Suffragettes tried to change the social situation and were very brave. The politicians didn't like the Suffragettes but finally, in 1918, women aged thirty had the vote and eleven years later in 1929 all women over 21 could vote.

THE NIGHTINGALE
AND THE ROSE

BEFORE YOU READ

1 Here are some words from the story. Do you know them? Match them with their pictures.

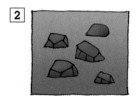

a. ☐ emeralds

b. ☐ coins

c. ☐ sundial

d. ☐ blood

e. ☐ tears

f. ☐ cart

g. ☐ wheel

h. ☐ mirror

i. ☐ oak

j. ☐ branch

k. ☐ moonlight

l. ☐ butterfly

m. ☐ daisy

n. ☐ Chamberlain

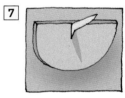

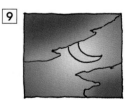

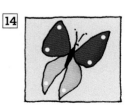

70

2 Write the correct word next to its definition. Then fill in the puzzle below to find out the secret message.

1. They are part of a tree. ....
2. It is a beautiful insect with big coloured wings. ....
3. It is a red liquid inside your body. ....
4. It is a very important party in a castle. ....
5. If you look into this object you see yourself. ....
6. It is a big tree. ....
7. It is a small flower with white petals and a yellow centre. ....
8. It is similar to a car but it has a horse and not an engine. ....
9. They are green precious stones. ....
10. He is a very important man in a city. ....
11. A car has four of these. ....
12. You can spend these objects in shops to buy things. ....
13. During the day there is sunshine and at night there is... ....
14. You can tell the time with this object if you don't have a watch. ....
15. When you cry these fall from your eyes. ....

1 ☐☐☐☐☐☐☐☐
2 ☐☐☐☐☐☐☐☐☐
3 ☐☐☐☐☐
4 ☐☐☐☐
5 ☐☐☐☐☐☐
6 ☐☐☐
7 ☐☐☐☐☐
8 ☐☐☐☐
9 ☐☐☐☐☐☐☐☐
10 ☐☐☐☐☐☐☐☐☐☐
11 ☐☐☐☐☐☐
12 ☐☐☐☐☐
13 ☐☐☐☐☐☐☐☐
14 ☐☐☐☐☐☐
15 ☐☐☐☐☐

PART ONE

The Student in Love

n a nest,[1] in a tree, in a garden a nightingale sang. Her song was beautiful: she sang of love and happiness. One day she saw a young student in the garden.

'She says she will dance with me if I give her a red rose,' said the Student, 'but in my garden there is no red rose.' And when he said this his eyes filled with tears. 'Every day I study philosophy and I read all the things that wise [2] men say about happiness. Now my happiness depends on a red rose!'

The Nightingale heard this and said to herself, 'Finally, here is a true lover. Every night I sing of love and this young man suffers for love.'

1. **nest** :
2. **wise** : informed; if you are wise you know a lot of things.

THE NIGHTINGALE
AND THE ROSE

The Student continued: 'I love the Professor's daughter and tomorrow there is a ball at the Prince's palace. My love will be there. If I take her a red rose she will dance with me. If I have no red rose she will not speak to me.' The young student looked around the garden. There were yellow roses and white roses but no red roses. 'Poor me,' he said. 'I need one red rose but I cannot see any in this garden.'

'Ah,' said the Nightingale, 'Love is a wonderful thing: it is more precious than emeralds, opals and pearls, you cannot buy love in the market place with gold coins.'

'The musicians will play their violins and my love will dance to their music. But she will not dance with me. I have no red rose to give her.' The Student fell onto the grass and started to cry.

A butterfly heard him and asked, 'Why is he crying?'

A daisy asked her friend, 'Why is he crying?'

'Why is he crying?' said a little green lizard. [1]

'He is crying for a red rose,' said the Nightingale.

'A red rose?' they all replied. 'That's ridiculous!'

The other animals laughed but the Nightingale understood. She watched the student sadly and thought of the mystery of love.

1. lizard :

UNDERSTANDING THE TEXT

 1 **Connect a word or phrase in each column to make sentences.**

1. The Student	will dance with the student	so	there are none in the garden
2. The Student	don't understand	if	he can't find a red rose
3. The Nightingale	is very sad	but	she sings a beautiful song
4. The Professor's daughter	thinks love is important	because	he gives her a red rose
5. The Student	is happy	but	now it's not useful to him
6. The animals	studies philosophy	so	finally she sees a true lover
7. The Nightingale	needs a red rose	because	they laugh at the student

1. ....
2. ....
3. ....
4. ....
5. ....
6. ....
7. ....

2 **Answer these questions.**

a. Why does the Student say 'My happiness depends on a red rose!'?

b. What does the Nightingale usually sing about?

c. What flowers are in the garden?

d. What does the Nightingale think of love?

e. Why does the Student start to cry?

f. What do the butterfly, the daisy and the lizard think of the Student?

PART TWO

The Nightingale's Sacrifice

In the middle of the garden there was a beautiful rose tree. The Nightingale flew to the rose tree and said, 'Give me a red rose,' she cried, 'and I will sing you my sweetest song.'

'I'm sorry, my roses are white like the snow on the mountain and the foam of the sea,' he answered. 'Ask my brother who grows round the sundial. Perhaps he can help you.'

The Nightingale flew to the sundial and said to the rose tree, 'Give me a red rose,' she cried, 'and I will sing you my sweetest song.'

'I'm sorry, my roses are yellow like the daffodil,' he answered. 'Ask my brother who grows under the Student's window. Perhaps he can help you.'

THE NIGHTINGALE AND THE ROSE

The Nightingale flew to the window and asked the rose, 'Give me a red rose,' she cried, 'and I will sing you my sweetest song.'

'I'm sorry, my roses are red like the coral in the sea but the winter was cold and my branches are broken. This year I have no flowers.'

'But I only need one red rose. Is there nothing I can do?'

'There is one thing you can do but I won't tell you. It is a terrible thing.'

'Tell me what it is. I am not afraid,' said the Nightingale.

'If you want a red rose you must build it by moonlight with music and colour it with your own blood. You must sing to me all night and press your heart against one of my thorns. All night you must sing and your blood will become my blood.'

'Death is a big price to pay for a rose,' said the Nightingale. 'Everybody likes life. I like life. I like to fly and to look at the flowers and to smell their perfumes in the wind. But love is better than life... and the heart of a man is much more important than the heart of a bird. The Student will have his rose.'

UNDERSTANDING THE TEXT

1 **William, the Student, and Emily, the Professor's daughter, are speaking outside her father's house. Read the dialogue and then complete William's notes.**

William: Well, Emily, will you come with me to the Prince's ball on Saturday?

Emily: I don't know, William. I will think about it.

William: Please, Emily, you know I love you. I will probably die if you don't come with me.

Emily: If I come with you will you give me a big present?

William: Of course. You know I will give you everything I have.

Emily: A ball at the Prince's palace is a very important occasion and I will wear my best dress.

William: Everybody will wear their best clothes, Emily.

Emily: Well I want a beautiful red rose to put on my dress. If you give me a red rose I will dance with you.

William: No problem. I will come to your house at 7.30 with the rose. The party starts at 8.00 and you will have the most beautiful red rose in the world.

Emily: Wonderful. Well, see you on Saturday, William.

William: Bye, Emily.

Prince's Ball

Day [1] ....

Time [2] ....

Place [3] ....

With [4] ..Emily.♡...

Clothes [5] ....

Present [6] ....

Where to meet [7] ....

What time to meet [8] ....

2 **Unscramble the questions and write answers.**

a. Where tree the rose white is?

....

....

b. What the sundial roses are round the colour?

....

....

c. Where grow rose red the tree does?

....

....

d. Why this flowers red does the rose have no year tree?

....

....

e. How a red Nightingale the make rose can?

....

....

f. What like do does to the Nightingale?

....

....

g. Is a the important more the heart a heart of man of bird than?

....

....

PET 3 **Look at the picture on page 83. With your partner describe the scene.**

a. Who is the boy in the picture?

....

....

b. What do you think he is feeling?

....

....

c. Why do you think he is feeling this way?

....

....

PART THREE

The Red Rose

he Nightingale flew back to the garden and saw the Student lying on the grass. His eyes were full of tears. 'Be happy,' the bird said. 'You will have your red rose and tomorrow night you will dance with your love at the Prince's ball. I will make the rose for you by moonlight, with music and with my own heart's blood. I ask you just one thing, you must promise to be a true lover.'

The Student looked up and listened but he didn't understand what the Nightingale was saying: he only understood things in books. But the oak tree understood and he said, 'Sing me your sweetest song, little Nightingale. I will be sad when you are not here.' The Nightingale sang for the oak tree. The Student heard the song and said, 'Yes, this music is very beautiful but can a bird really understand love? She sings well but she is like an

THE NIGHTINGALE AND THE ROSE

artist and everybody knows that artists are not sincere. She thinks only of music and could never do anything practical to help anybody.' He got up, went into his house, lay on his bed and slept.

When night came and the moon shone, the Nightingale flew to the rose tree. She pressed her heart against one of his thorns. All night she sang her sweetest songs. The cold crystal moon listened and the Nightingale's blood slowly left her. At the top of the rose tree a flower started to grow. First it was pale; silver like the new day. But the tree cried 'Come closer!'

The Nightingale came closer and sang louder, then the rose became pink like a red rose in a silver mirror.

'Come closer, little Nightingale,' said the rose bush. 'Come closer. If not, the day will come before the rose is finished.' The Nightingale came closer and as the thorn pierced [1] her heart she sang of a love that never dies. She felt a strong pain and her voice became softer and softer. Finally the rose was ready, a marvellous red rose, red like the eastern skies.

Then the little Nightingale sang her most beautiful final song. The white moon heard it and she forgot the sun in the East and stayed in the sky to listen. The red rose heard the song and opened her petals in the cold morning air. The sleeping shepherds woke up when they heard it and the river carried its message to the sea. The rose tree heard the song and cried, 'Look, little Nightingale, look. The rose is finished.'

But the Nightingale didn't hear because she was dead on the grass with the thorn in her heart.

1. **pierced** : went into.

UNDERSTANDING THE TEXT

1 **Answer the following questions.**

    a. What did the Nightingale want from the Student?

    b. What did the Student think of artists?

    c. How did the Nightingale make the red rose?

    d. What did the rose bush say to the Nightingale?

    e. Who heard the Nightingale's song?

    f. Why didn't the Nightingale hear the rose tree?

PET **2** **A.** **Listen to William and then answer the questions. Choose the best answer A, B or C.**

1. William is
- [ ] **A** a student
- [ ] **B** a professor
- [ ] **C** a philosopher

2. He loves studying
- [ ] **A** philosophy and art
- [ ] **B** history and Italian
- [ ] **C** philosophy and history

3. His mother and father
- [ ] **A** live in a small house
- [ ] **B** buy him a lot of things
- [ ] **C** are very beautiful

4. Emily
- [ ] **A** loves William
- [ ] **B** is the Professor's daughter
- [ ] **C** loves all the boys

5. Emily will dance with William if
- [ ] **A** he buys her lots of things
- [ ] **B** he stops studying
- [ ] **C** he gives her a red rose

B. **Listen again and answer these questions.**

    1. What are the two things that William loves?

    2. What is William's problem?

    3. Is it difficult to find a red rose?

Now listen to Emily and answer the questions. Choose the best answer A, B or C.

1. Everybody says
    - [ ] **A** Emily loves all the boys
    - [ ] **B** Emily's mother is a professor
    - [ ] **C** Emily is very beautiful

2. The artist's name is
    - [ ] **A** Jonathan
    - [ ] **B** Richard
    - [ ] **C** William

3. Emily's favourite boy is
    - [ ] **A** Richard
    - [ ] **B** Tristram
    - [ ] **C** William

4. Emily likes boys who
    - [ ] **A** dance well
    - [ ] **B** think she's beautiful
    - [ ] **C** spend a lot of money

5. Who will she dance with at the Prince's ball?
    - [ ] **A** William
    - [ ] **B** Tristram
    - [ ] **C** Jonathan

4 **The Nightingale thinks that love is very important. She dies for love. But is love more important than other things?**

A. **Look at these words and put them in order. The most important is number 1 and the least important number 10.**

- [ ] love
- [ ] friends
- [ ] family
- [ ] fun
- [ ] music
- [ ] beauty
- [ ] peace
- [ ] money
- [ ] health
- [ ] food

Speak with your teacher and your class about your choice. Do you all agree?

B. **What did you put as number 1? Write a sentence in English to explain why you chose this.**

PART FOUR

The Professor's Daughter

he next day, at lunchtime, the Student woke up and looked out of his window. 'That's lucky,' he said, 'here is a red rose. It is an extremely beautiful red rose. I'm sure it has a long Latin name.' He took the rose from the tree. He put on his hat and ran to the Professor's house. The Professor's daughter was sitting near the door.

'Look, here is a red rose for you. Tonight you must dance with me as you promised. You will wear it next to your heart and I will say "I love you." '

The girl didn't smile but she looked at him. 'I'm sorry,' she said, 'I don't like the colour. My dress is blue and the rose is red. And another thing, the Chamberlain's son gave me jewels.

THE NIGHTINGALE
AND THE ROSE

Everybody knows that jewels are more expensive than flowers. I don't want your rose.'

'You are very ungrateful,' said the Student angrily, and he threw the rose into the street. At that moment a cart passed and the wheels crushed the flower.

'You are very rude,' said the girl. 'I will dance with the Chamberlain's son, not with you.' Then she stood up and went into her house.

The Student started to walk home. 'Love is a stupid thing,' he said. 'I prefer to study books. They are much more interesting and useful... Yes, logic is much more useful then love. I will go home and study philosophy and metaphysics.' And that's what he did.

UNDERSTANDING THE TEXT

1 **Answer these questions about the story.**

a. Why does the Student want to find a rose?

....

....

b. What does the Nightingale think of love?

....

....

c. Who says 'That's ridiculous'? Why?

....

....

d. Where is the white rose tree?

....

....

e. Where is the yellow rose tree?

....

....

f. Who tells the Nightingale how to get a red rose?

....

....

g. What must she do?

....

....

h. Why is the student crying?

....

....

i. Is the Professor's daughter happy with the rose? Why? Why not?

....

....

j. What happens to the red rose?

....

....

GRAMMAR

> **Will and *if***
>
> **Will** indicates that an action happens in the future. It is always followed by the infinitive without **to** and it doesn't change form in the third person singular.
>
> Look at these examples:
>
> *The young King **will be** crowned tomorrow.*   **will + infinitive** (be)
>
> *Perhaps we **will find** a pot of gold.*   **will + infinitive** (find)
>
> *I **will make** a red rose for you.*   **will + infinitive** (make)

1 Choose the correct verb and complete these sentences with *will*.

> be   buy   visit   go   do

a. We .... to Spain on holiday next year.

b. I .... all my homework before I watch TV.

c. Andrew .... very happy when the exam is finished.

d. They .... a new car when they have enough money.

e. Susan .... me tomorrow.

> Sometimes the verb is negative. **Will + not** is often written as **won't**.
>
> *'I **will not listen** to you,' says the Star-Child.*
>
> *'I **won't wear** the crown,' says the young King.*

2 Write two things you *will do* tomorrow and two things you *won't do* tomorrow. You can use some of these verbs if you want:

> go   do   eat   watch   sleep

a. ....

b. ....

c. ....

d. ....

90

Now look at these examples:

*The student says, 'If I **take** her a red rose, she **will dance** with me.'*

*'If I **don't take** her a red rose, she **will not dance** with me.'*

If + present , will (not) + infinitive

 3 Connect a line in column A to column B and to column C.

A	B	C
1. The Student says,	'If we leave the baby here,	I will sing my sweetest song.'
2. The Nightingale says,	'If you don't give me some money,	she will dance with me.'
3. The Magician says,	'If you give me a red rose,	he will die.'
4. The old man says,	'If you find the gold,	you will be free.'
5. The woodcutter says,	'If I take her a red rose,	I will die of hunger.'

 4 Complete these sentences with your own ideas.
Remember *If* + present, *will (not)* + infinitive.

   a. If it rains on Sunday, I will ....
   b. I will dance with you if you ....
   c. If I eat too much chocolate, ...
   d. If I find a lot of money, I ....
   e. My mother will be very angry if I ....
   f. If I don't do my homework, my teacher ....

T: GRADE 5

 5 **Topic – Future Plans**
Tell the class about your plans for the future. Use the following questions to help you.

   a. What are you studying at the moment?
   b. What hobbies and interests do you have?
   c. What job you think you'll do in the future? Why?
   d. Do you think you will change your hobbies? Why?

Similes

Similes help to describe an object or a person. They often use the word *like*. Oscar Wilde used a lot of similes to write beautiful descriptions.

What colour is the Star-Child's skin?

Ivory is always very white.

The Star-Child's skin was **white like ivory**.

 6 Look at some more examples from the stories. Connect the three columns A, B and C to make complete sentences. Check your answers in the stories.

A	B	C
1. The master of the boat was	blue	like the snow on the mountain
2. The pearl was	white	like violets near a river
3. His hair was	round and white	like daffodils
4. His eyes were	gold	like the new day
5. His skin was	white	like the moon
6. The rose was	black	like a red rose in a silver mirror
7. The rose was	pink	like ivory
8. My roses are	silver	like ebony

7 Now you try.

a. Round
   The pizza was round like the .... (think of something always very round)

b. Tall
   The .... was tall like a .... (think of something very very tall)

c. Silver ....

d. Beautiful ....

92

EXIT TEST

COMPARING THE STORIES

1 **Opposites**
Connect each adjective with its opposite. Translate the words into your language.

a. sad		**1.** beautiful
b. humble		**2.** rude
c. poor		**3.** intelligent
d. polite		**4.** interesting
e. ugly		**5.** rich
f. boring		**6.** happy
g. stupid		**7.** arrogant

2 Connect each adjective with its definition. Translate the adjectives into your language.

> lazy   sincere   elegant   greedy   helpful   in love

a. A person who wears fine clothes is ....

b. A person who wants more than is necessary is ....

c. A person who likes to help people is ....

d. A person who loves another person is ....

e. A person who doesn't like working is ....

f. A person who says the things they think is ....

3 In all of these stories by Oscar Wilde the main character is a boy or young man. The three characters change during the stories. Choose from the adjectives on page 93 and describe them. You can use all of the adjectives more than once.

The young King is ....
at the beginning of the story and ....
at the end of the story.

The Star-Child is ....
at the beginning of the story and ....
at the end of the story.

The Student is ....
at the beginning of the story and ....
at the end of the story.

4 Which of the characters do you like best? Why?

....

5 Which of the characters don't you like very much? Why?

....

6 In two of the stories not only humans speak.

a. Which stories are they and who speaks?

b. Do you like the idea of animals and plants that speak in stories? Why? / Why not?

c. There are many speaking animals in traditional stories. With your partner try to think of as many as possible in three minutes.

7 In two of the stories young boys become kings in unusual situations.

    a. Which stories are they?

    b. What is unusual about the way they become king?

    c. In many stories and films children or normal people become princes or princesses. How many can you think of?

8 Oscar Wilde's stories have a lot of children, magic and surprising events. Do you think they are good stories for young children to read? Why? / Why not?

9 You can find these elements in the three stories in this book. Connect them to the correct title. Sometimes you must connect the word to more than one title.

> love   hate   death   suffering   poor people   angry people
> magical clothes   beautiful jewels   speaking animals
> a poor person becomes rich   a child who doesn't know his father
> sacrifice   one person transforms into a different person
> a king   a queen   a beautiful girl

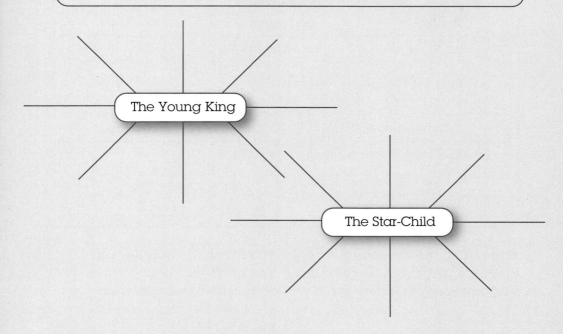

The Young King

The Star-Child

95

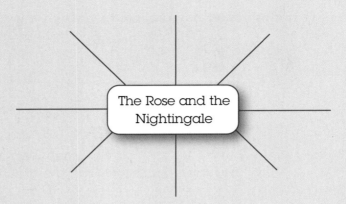

The Rose and the Nightingale

Check your ideas with your partner. Do you agree? Which of these elements are usually in stories for children? Which of these elements are not usually in stories for children?

10 Write a story for children.
In a small group try to think of some different ideas to complete the table? Think of books, films, comics etc.

Typical good characters	Typical bad characters	Typical place
princess, poor child	witch, wolf	forest, castle
....	....	....
....	....	....

Typical events		Typical ending
find treasure, meet a ghost		....
....		....
....		....

Write all of the ideas on the board. In your group choose your favourite ideas and invent a story. You can talk about your ideas in your language but then try to write it together in English.